CONSTRUCTIVE
BEEKEEPING

BY
ED H. CLARK

CONSTRUCTIVE
BEEKEEPING

BY
ED H. CLARK

W

A WORD TO THE BEEKEEPERS.

For this system or theory of housing bees I am making no extravagant claims, and had no thought of giving the public the result of my study and experiments until I had given them a test of a few more years, but the present necessity for increased production makes it the first duty of every person to stimulate in every way that he can this production; and if this system proves to be that will-of-the-wisp that the beekeepers have been chasing, a conclusion will be reached sooner than could be by my individual effort.

—ED. H. CLARK.

CONSTRUCTIVE BEEKEEPING

Commercial beekeeping has for its object the production of the maximum quantity of well ripened honey at a minimum of cost. Swarming adds greatly to the cost of producing honey. Most methods of swarm-prevention have in them the element of destructiveness. Ventilating, removing the queen, shaking the bees, removing the brood, exchanging brood-bodies, loosening the cover, all destroying something that the bees have done, or adding to the work to be done in the hive.

Let me state here that you are not going to be told that absolute swarm-prevention is a possibility, nor will you be told that honey can be produced by absent treatment. What you will get from a good understanding of the following pages is that the beekeeper who takes advantage of the laws relating to condensation of vapor, and follows where the bees have been leading will have advanced one step nearer the swarmless bee.

Instead of going to the hive and telling the bees (by manipulations) "don't do this"; say to them "keep all your brood, keep your queen, keep the cold damp air out of the hive and I will make your hive so perfect a condenser of water vapor that the work of evaporating water from the nectar will be done quickly." We will bring team-work into play, and each get the benefit of every advantage gained.

Constructive beekeeping helps by getting the honey ripened quickly each night and stored out of the way of the queen. The hive will then be maintained in such a condition that the bees have the greatest amount of comfort in relation to the results produced.

EFFICIENCY

Familiarity with the bees makes a person wonder where the mule got his reputation for stubborness. We must, when we work with bees, follow where they lead if we want satisfactory results.

They generally know what they want and fight to do it in their own way.

The bee is an economical "fellow." Efficiency is his last name. For ages before our time he has been working at the same task that he has brought to such perfection. The bees that have excited the amazement and wonder of the world as builders of geometric comb, with all its economy and strength, the bees that make from the nectar of the flowers the honey, that is the superlative of all things sweet and good; have not stopped with these two achievements, but have maintained a marvel within their home, which, when understood, will amaze us more than ever.

The bees collect propolis (despised by the beekeeper) and with infinite work and a knowledge of the laws of condensation cf moisture, have varnished the inside, sealed all openings that would give upper ventilation, making out of their home a perfect condenser for the water vapor that has filled the hive air by evaporation from the nectar. This moisture they collect from the inside surface of the hive, works right in with their idea of economy, saving the carrying of water from the stream or spring.

CO-OPERATION

One great discovery in beekeeping was the beespace. When this space was within the limits that conform to what instinct told them was best, peace between the beekeeper and his bees was established along this line.

A similar misunderstanding is on at the present time about the best way to get rid of the moisture in the hive. The bees, from the habits and customs, carried down for ages, contending that condensation is the system best adapted to their mode of life, while the beekeeper is trying to force them to use the ventilating system. I think that the beekeepr, being a most sensible person, will see the error of his way and eventually follow where the bee leads. The bees but show their contempt for ventilation when it is at its best.

URGE

Swarming is the breaking in two of a populous colony of bees for the reproduction of the colony. Other swarming need not be considered here.

The confusion of conceptions which arises in the mind when the term swarm-prevention is mentioned, has to a great extent, been a hindrance to logical reasoning on this subject. At mention of the word swarm, there is conjured up from the background of our minds, the exciting and spectacular scenes we have seen enacted around the aipary. What we want to prevent not the birth but the conception. We must go even prior to this and prevent the primal urge. To avoid confusion and give reason and logic an opportunity to work, unhampered by visions conjured up at mention of the word swarm, the word urge will be used here to refer to the initial starting point of the swarming impulse.

The urge is the direct result of the actual or apparent failure of the queen.

With the exception of a young queen in the hive, all other methods of urge prevention have to do with room for the queen to lay. Give the queen plenty of room to lay and swarms are reduced. Give the colony plenty of ventilation and you have fewer swarms. Young queen, and shade in the afternoons, reduce swarming. Room, ventilation, shade in the afternoon, and young queen have given the best results so far obtained.

CONDITIONS.

What has puzzled the beekeeper is that results obtained by the same treatment year after year have not been the same. Nor have the same results been obtained by similar treatments in different localities. Bees always respond to similar treatment under like conditions by giving uniform results. The beekeepers' trouble is that he makes his treatments uniform but not his conditions. Conditions include all of the following things:

1. Temperature.
2. Direction and velocity of the wind.

3. Moisture in the air.
4. Depth of evaporation.
5. Sunshine.
6. Altitude.
7. Number of bees.
8. Fertility and vigor of the queen.
9. Number of worker cells in the hive.
10. Size of entrance.
11. Construction of hive.
12. Fit of hive parts and cover.
13. Radiating surface on inside walls of the hive.
14. Number of honey and pollen-producing plants.
15. The amount of water in the nectar.

To show how much harder bees would have to fan at the entrance to ripen honey in some localities than in others, if they ripened it altogether by ventilation, the depth of evaporation in inches, taken from a map of the Weather Bureau, is here given:

25 inches, Ontario, Canada.
35 inches, Massachusetts.
30 inches, Ohio, Northern part.
50 inches, Illinois, Southern part.
50 inches, Texas, Central part.
60 to 100 inches, Colorado, New Mexico, Arizona and most of California.

From this will readily be seen the different conditions under which the bees do their work in different localities, due to only one of the conditions previously mentioned. It is apparent that similar treatments, as regards ventilation, cannot give uniform results in all these places.

To make the treatments as near uniform as possible and get like results under these different atmospheric conditions, we must follow the way of the bees and make the hive a condenser instead of a ventilator.

ROOM

To look at the question of room and apply constructive treatment is our aim. If the hive room is large, good results do not always follow. So much thought has been put into the experiments carried out with different sizes of hives, that the majority of beekeepers have settled on a standard size, which is best under most conditions. Some advocate one a little larger, some a little smaller, but neither deviate much from the recognized standard. So we must conclude that the bees and their keeper do not disagree greatly on the size of the hive.

During the urge period the attention of the beekeeper has mostly been given to brood rearing. To get best results he must have the hive filled with bees at the beginning of the honey flow. He knows that swarming results when the hive is full of brood, nectar, honey and bees, so he increases the size of the hive.

As the lack of room is not the result of one condition, but of many, the added room may be more than the bees can heat. Our aim then must be to determine that nice balance between too much and too little room, ventilation and shade. On the correct determination of this balance our success depends.

To have in our minds a conception of brood conditions in the hive prior to and at the time of the urge, will give us a clearer understanding of the principal cause of the urge. By a study of the following hypothetical tabulation, the conditions of a hive can be seen. A ten-frame hive of Langstroth dimentions has, when the frames are filled with worker cells, about 66,000 of these cells. Deduct 12,000 for the ones occupied by pollen and capped honey and there is left 54,000 cells for brood and nectar.

This is about the condition of brood in a strong colony at Fargo, North Dakota on the given dates. On June 23 there would be in this colony 25,000 bees under 10 days and 44,440 over 10 days old.

It can readily be seen that, when the hive is filled with brood honey and pollen, when sunshine and flowers are tempting

APRIL	EGGS	BROOD		EGGS	BROOD
1	100		13	1500	24100
2	100		14	1200	24900
3	110		15	1000	25400
4	110		16	1600	26450
5	115		17	1600	27400
6	115		18	1800	28500
7	120		19	1500	29200
8	120		20	1600	29900
9	120		21	1800	30700
10	140		22	2200	31800
11	140		23	2400	33000
12	140		24	300	34800
13	180		25	2800	36360
14	180		26	2400	37300
15	200		27	2600	38500
16	200		28	3000	40060
17	210		29	2500	41000
18	215		30	2400	41900
19	220		31	2000	42300
20	240		JUNE.		
21	300	3375	1	2400	43000
22	300	3575	2	3000	44300
23	400	3875	3	3400	46200
24	500	4265	4	3000	48000
25	550	4705	5	3200	50200
26	650	5240	6	3100	51700
27	700	5825	7	3000	53100
28	800	6505	8	2400	53700
29	900	7285	9	3000	55200
30	1000	8165	10	3200	56800
May			11	3000	58000
1	1100	9125	12	2500	58300
2	1200	10185	13	2400	58300
3	1200	11245	14	3000	58300
4	1300	12365	15	2800	58300
5	1400	13585	16	2400	58300
6	1400	14785	17	2600	58300
7	1500	16085	18	3000	58300
8	1500	17375	19	2500	58300
9	1500	18660	20	2400	58300
10	1600	20040	21	2000	58300
11	1700	21500	22	2400	58300
12	1700	22900	23	3000	58300

the bees to work almost to madness, the honey coming in deposited in cells and interfering with the laying of the queen, their system of housekeeping is disorganized. Their instinct tells them that now is the time to get things in shape for swarming.

Another condition of hive that gives like results is when the cells are filled with nectar, and the conditions of atmosphere and hive such that little or no evaporation takes place.

WAX.

Also, when from whatever cause, the bees have a lot of sealed honey around the brood-nest, and because the empty outside combs are too cold for work, they will not remove the honey to these to enable the queen to lay. This is the condition of colony where spreading the brood makes more room, also that makes a colony in a sixteen-frame hive take on the urge.

To understand this condition a knowledge of the relation of temperature to wax-working is necessary. To work and build with wax requires a temperature nearly as high as that maintained in the brood-nest, and when any part of the hive is colder than is consistent for easy work with wax; the bees, if they work on wax at all, do it in the warmer part of the hive.

It must be borne in mind that the bees, like all other creatures, do things mostly in the line of least resistance. It is easier to work wax where it is warm than where it is cold. Hence we find the bees starting on the center combs in the second super, before finishing the outside ones in the first. The beekeeper who puts boards, with slits on either side for the bees to enter the super from the brood-chamber, produces a condition that sends the warm air from the brood-chamber up through the slits; thereby giving the bees a good temperature to work on outside combs.

Beekeepers who produce heather honey in England, because of the cold nights, put a covering over and around the supers to keep them warm. Otherwise poor results follow. It has been demonstrated that combs can be built down to the bottom-bar in a super better than in a brood-chamber with an opening at the bottom front, and here again temperature plays its part.

Knowing the relation of temperature to working in wax, it can readily be seen how a colony, with plenty of hive room may be crowded for brood-room and acquire the urge.

Caging the queen gives more room, but brings on the urge and may result in swarming. Removal of the brood and substituting foundation or drawn comb gives laying room to the queen. These are good examples of destrutive beekeeping. How much better the colony could get ahead had they hives so constructed and protected from outside weather conditions that they could evaporate the water from their honey and store it away from the brood-nest.

Shaking the bees gives ample room, but that is artifical swarming.

Removing the water from the nectar and giving the bees an opportunity to store this ripened nectar in fewer cells makes more room, and when this can be accomplished soon after the bees' last trip to the field, it shortens their hours of work, adds to their comfort and perhaps lengthening their lives.

It is just as necessary for the economy of their domestic arrangement that they have room sufficient and warm enough to ripen the honey, as for storage of ripened honey and brood.

BROOD - FRAMES

Brood-frames cannot be ignored in constructive beekeeping, because they have so close a relation to room in the hive. The nearer the number of worker cells in all the frames approach the maximum, the more room the hive contains. Frames not filled to the bottom-bar and frames containing drone comb; will, together with other conditions, make just enough difference in hive room to bring on the urge. The ten-frame hive, under these conditions, has but little more room than eight full frames.

The size of a hive, no matter what its dimensions, is expressed in the number of worker cells it contains. A hive with but from 40,000 to 50,000 worker cells must not be thought of as a standard hive, even though it has ten standard frames.

Empty spaces inside the frames make an extra expense to the beekeeper, and an added burden on the bees. They have fully one-fourth more space to keep warm in a hive with beespace and frames, than they have in a box hive of the same dimensions.

Another way a constructive principle can be applied to the frame is to reduce the top-bar to the same thickness as the bottom-bar. Beekeepers are convinced that ten frames are not enough for a good laying queen, therefore some advocate giving more room above, by adding another hive-body, others giving more room on the side, using a wider hive-body, while some make the hive higher, wider and longer. The larger hive-body, it is claimed, gives better results when used in producing extracted honey, but it can never become a standard hive, because a great number of beekeepers produce comb honey.

The ten frame or standard hive being the best for most purposes, it remains for the beekeeper, when he gives more room, to give the bees a brood-nest with as little space between comb and comb, where the brood-bodies meet, as possible. Considering that all the combs in the hive body given above are built down to the bottom-bar, we have at the junction of the two hive bodies, a space between comb and comb of top-bar, beespace and bottom bar, or about one and one-half inches. From experiments it has been observed that a thick top-bar is a fairly good queen excluder. Hence it is apparent that more or less of an effort is required to pass over this space. When this is reduced by a thin top-bar an easier passage for the queen and bees is made, and better results will follow. In smoothing out little difficulties like these we are always amply repaid.

For the same reason all comb should be built down to the bottom-bar, then with cells overlapping top and bottom bars of the respective hive-bodies, the beespace is all that separates comb from comb. The thin top-bar adds more cells to the hive, so adding its mite to the economy of the hive, and giving the bees nearly continuous comb from top to bottom of the hive.

VENTILATION

One of the axioms of beekeeping is ventilate .   No matter what other manipulations the beekeeper recommends to prevent swarming, he always adds "VENTILATE".   Why do the bees require so much fresh air in May and June when they have double the capacity of air space, within the hive, that they   have   in October and November

The air in the hive serves a two-fold purpose.   It supplies the bees with oxygen and influences the amount of water vapor the hive can contain.   This water vapor is given off in the ripening of honey and the breathing of the bees.   It is either removed from the hive by ventilation or carried to the walls and ceiling, where it is condensed and used by the bees for domestic purposes.

Ventilation is constructive in principle when the nights are warm and the atmosphere comparatively dry, but is destructive when the nights are cold and the atmosphere nearly saturated with moisture. Only the best beekeepers can, by ventilation, maintain that nice balance that accords with temperature and strength of the colony.

During the season when the bees are most likely to acquire the urge the nights are colder and the humidity of the air greater than later in the season.   At such times a cold current of air throught the hive interferes with the rearing of brood and does not help much in the ripening of honey.   Each day of good honey gathering bringing the colony nearer to a crowded condition.

No beekeeper finds it practicable to change the ventilation as the temperature changes, but must rely on his judgment of the amount that gives best results.

The manipulations of divisible brood chambers can be considered under the head of ventilation.   When the two sections of the brood chamber have been broken apart or exchanged, the percentage of colonies that acquire the urge has been reduced. This can readily be explained by showing how it makes for ventilation.   If it is done prior to the urge it will have a tendency to reduce swarming. All tearing apart of hive-bodies from each other or from covers or bottom-boards leaves a rough surface to the adjoining parts, the result of dirt, grass or bees adhering to the propolis. When the hive is again put together the adjoining surfaces are not tight and ventilation is the result.

The result of ventilation is room. Because ventilation has given more room we have all believed in it. Room can be given by ventilation but at the expense of a warm hive.

The velocity of the wind and shape and size of the entrance must be studied in their relations to one another when considering ventilation. The ill effects from winds, even with reduced entrances, is sometimes apparent. A bunch of grass or a shrub, sometimes two or three feet away from the entrance may deflect and send a current of air right into the hive, or it may produce a calm at the entrance. The only possible way to have uniform ventilation is to have a good wind break. If the hives are so situated as to be in a calm at all times then what ventilation there is is regular, but hives so situated are troubled with excessive swarming.

With any wind at all blowing, no two hives in a row, nor in different rows, even though they have the same size openings, will have the same ventilation. To any one who has tried to follow the course of air movements along the surface and a few feet above the earth, winding in and out among obstructions, sometimes flying off at a tangent, sometimes circling all around, sometimes upward or downward, bounding and rebounding from such obstructions, it is apparent that air currents among the hives in an aipary are so complex as to be past our understanding. This but shows with what little certainty ventilation can be made anyway uniform. Hence the different results at different times from apparently similar treatments.

Shade in the afternoon has been proven to be somewhat of a preventive of the urge. This is clearly explained by the laws governing condensation of water vapor, and will be treated in the chapter under that head.

Ventilation and shade each make more room in the hive,

which leads me to remark that ventilation is constructive at the proper time, and under right atmospheric conditions, but is destructive in spring and early summer, because of the low temperature and moisture-laden atmosphere. Air that is already full of moisture will be of little use in drying the air of the hive. To demonstrate this a wet cloth can be hung by the hive on a night of high relative humidity, and in the morning it will be far from dry. It will require no great reasoning on the part of anyone to see that the water vapor of the hive is not all carried away by ventilation.

SWARMING

During a heavy flow of honey the queen gradually and sometimes sharply reduces the number of eggs laid per day. We know that had the same thing happened during the spring building-up of the colony it would have brought on the urge. Some one will point to the above statement as a contradiction of the one that the urge is the direct result of the actual or apparent failure of the queen. We know, from observation, that they are almost sure to swarm in spring or early summer when this condition exists. In summer, with an abundant honey flow and plenty of room, they rarely swarm. In the fall they quietly supercede the queen.

It is obvious that the aim or object for which they strive in each of the three periods mentioned is not the same, and that the aim or object attained as the collective result of these three periods, is that for which all bees are striving, the survival of the colony.

During the spring period the one purpose for which they strive is numbers. Every atom of concerted energy of the colony is working to increase the unit strength of the colony. When we consider the advantages that numbers give for offense and defense and general welfare, we know that they are urged on by the law of the survival of the fittest.

After this carnival of brood-rearing, when weather and field conditions prompt their instinct to a realization that winter is ahead, and only the colonies that are fitest survive; we have the energy of the colony directed as a unit to gathering stores. The brood-rearing at this period is a secondary consideration, and about keeps pace with the bees that fall from the ranks.

The next period has brood-rearing and general preparation for the winter on its schedule. This brood-rearing does not take on the riotous nature of the spring period.

There is no definite line or date dividing these periods. They dovetail into one another and only by close observance and much bee-wisdom is a person able to detect the change of colony procedure as the bees pass from the spring to the summer period. Because of the inability of the ordinary beekeeper to sense this, he has not been uniformly successful in the production of comb honey.

In passing from one period to the other the colony work fades from rearing brood to gathering honey, then back again to rearing brood. This change depends as much or more on strength and good condition of the colony as on weather and field conditions. The beekeeper, not being able to tell the exact time that each particular colony is prepared to take on the work of the next period, assumes that they take on the urge in the summer period when contraction of the brood takes place.

The rearing of queens by the colony cannot be considered an acquired habit. We have to admit that it is both for the reproduction of the colony and the species, and goes back to the very origin of life.

The strongest disire or urge in all life is to reproduce its kind. If this disire is destroyed in a species that species becomes extinct. It is an impossibility to breed a swarmless bee. If it could be done all ambition, vim and energy would disappear, and as much could be accomplished with so many inanimate units.' The main thing in life is reproduction. All other things are done for the better bringing it about. Increase and multiply is the urge of all things living, so that each generation is greater in numbers and wisdom than the preceding.

As generation after generation succeeds one another, the

acquirements of one when they are advantageous to the next, puts that next generation a stride forward in the battle for ex-istance . The colonies in which this acquired habit took hold suc-ceeded best, and when we realize that this development has been going on for countless years, by slow evolution, we can believe that the bees have acquired habits that are of advantage to them in their mode of life. Swarming is one of these habits, and should we be able to breed it out, they would be thrown back to their way of life or social development that existed in the colony before they acquired this habit.

A colony of bees cannot be considered in the scheme of re-production the same as individuals. Individuals are being born regularly in th ecology, but through an acquired habit the colony carries on what may be called limited birth control, by prevent-ing the perfect development of the female. CONDITIONS that induce the colony to start queen-cells, or allow perfect development, is what we must control to prevent the urge.

The bees in their wild state seldom had a home that did not at some time become filled to its capacity. They developed a habit, as we know, of putting off reproduction of the colony while conditions in the home were such, that the strength of the colony could be increased; but letting it take place normally when home conditions were getting crowded. Sometimes the home was of such size that this did not occur for several seasons, but eventually the home became crowded, from accumulated honey, and swarming resulted.

The smaller the hive the sooner the urge. The smaller col-onies that came from the smaller homes did not survive as often as the stronger ones, that had more time and room to prepare; hence we have bees with the habit of putting off swarming as long as hive conditions permit.

The act of swarming must be considered a protest against a condition in the hive. Could we keep a colony of bees with a good queen, and plenty of necessary room, we would have no swarms. It has been done, so it is not impossible.

EVAPORATION

One of the by-products given off by the bees in ripening honey is water vapor. Sometimes more than one pound of water is expelled from the honey in part of a night. This much water vapor at average hive temperature of 90 deg. F. would fill to saturation over 600 standard hive-bodies. Then if the air entering the hive was absolutely dry, and that emerging from it at saturation, and the hive air completely changed in one minute, this amount of water vapor would pass out in about 10 hours. But we never have absolutely dry air, nor do we have all the air in the hive even near saturation, nor can the bees change all the air in the hive, through an opening in the lower front, in one minute. The bees could not get rid of a pound of water this way in 24 hours.

One pound of water evaporated during the night, from the nectar brought in during the day, would empty over 1600 cells. These and the ones vacated by hatching bees would keep well ahead of the queen.

In handling bees we are bound to give recognition and allowance for the storing and ripening of nectar between the stores of ripened honey and the brood. If the hive be so protected that a good working temperature can easily be maintained, the honey close around the brood-nest is either moved to some outside comb, or consumed by the bees and brood, leaving empty cells to store the nectar.

Removal of the water from the nectar is something to which the beekeeper has given little attention, because it could not be seen, but nevertheless we have the water vapor given off when the bee ripens a drop of honey in his mouth. We are as certain of it as we are of the wax scales that the bees secrete. We see the wax, but the water vapor not being visible, has claimed little of our attention. The beekeeper, striving for good results, cannot afford to overlook the great part that water vapor plays in the hive.

When the beekeeper opens the hive for inspection, does he

ever observe the room available for the storage of a generous amount of nectar. He looks for queen-cells, the amount of brood and sealed honey, and in this way he is conscious of what he calls plenty of room. While there is yet space between brood and honey, he concludes that the bees swarm before they are crowded for room. Did they have room to carry on all the work of the hive?

Consider a farmer, who has ample granary room to store his wheat, and who tried to put it in the granary before it was thrashed. It can readily be seen that the granary would not hold it. The bees must have temporary storage cells just as the farmer must have sufficient room to stack his grain before it is thrashed. This is the room that the beekeeper has sensed, rather than to have observed it as room necessary for the domestic work of the bees. As the farmer must put aside the straw, so must the bees evaporate the water to get the finished honey that can be stored economically.

How do the bees dispose of the great amount of water carried into the hive in the nectar? You say: "they evaporate it and it is carried off by the air." But is it? Try drying the family wash in a room of dimensions, as regards water to be evaporated, temperature and opening at the floor, proportionate to the beehive, and see how fast your wash dries. Be sure you have no windows in the room to condense the water vapor. Try it when temperature outside is 60 deg. F and the atmosphere one-half saturated, (one-half saturated in the evening is about as dry as we find the air in May and June where there is ample rain fall). Then try it when the atmosphere is 9-10 saturated, and after this experience you will be more amazed than ever at the bees' efficiency.

Still you are skeptical and remember that you should have put an electric fan at the opening, because the bees are seen to fan at the entrance. Let us not delude ourselves about the bees moving all the water vapor, given off in ripening honey, out of the hive by fanning.

Records of temperature and humidity taken from the U. S.

Weather Station at Moorhead, Minnesota, for three days at 7 P. M. are as follows:

1916	Temperature	Humidity
May 13	45.5	93
May 14	42.5	100
May 15	37.5	96

To remove from the hive one grain of water vapor by fanning on May 13, the volume of air that would have to be moved would be five or six times the air capacity of the hive. To remove one pound of water would require the removal of a volume of air equal to the capacity of from 30,000 to 40,000 hive-bodies. On May 14 the outside air being saturated, no water vapor could be removed by a change of air, and on May 15 the result would be about the same.

From this it can readily be seen that little evaporation from the nectar could take place by ventilation, and we are led to believe that on such days as these the urge takes hold, giving swarms the last week of May.

No contention will be raised when the statement is made that the water is evaporated from the honey before we have ripened honey. It has been taken for granted that as soon as the water passes off by evaporation the bees were done with it. They would be if this evaporation took place the same as from the family wash hung on a line in open air. But if instead of having the great volume of mooving outside air, we have in a standard hive body a little less than one cubic foot of air, where is all this water vapor going to go? Saturate this small volume of air at hive temperature and the water vapor in it would be from 1-600 to 1-5000 of what the bees evaporate in a night during a good honey flow. If they saturated this air in the hive and then forced it out at the entrance, they would cause rain in the hive near the entrance.

When hive conditions are as they should be we must conclude that all the air in the hive is never near saturation. There is no known way of accurately measuring the humidity of the air inside the hive, but we can safely estimate that, except for a thin layer under the cover, it is never more than 3-4 saturated. We have then air entering the hive at a temperature of 50'F. and 1-2 satuation which would, when heated to hive temperature, be about 1-4 saturated. This air then when heated to hive temperature, being 1-4 saturated, is capable of taking up moisture until it is 3-4 saturated. The amount of moisture taken up by this air is 3-4 minus 1-4 or 1-2 of its capacity at saturation. This amount of water vapor would be less than 1-1000 of a pound for one cubic foot of air.

Now follow the course of this 90 degree air in passing from the hive by way of the entrance. As it approaches the entrance the temperature is reduced and it would give up some of its water at the entrance. Could they force all the air out of the hive at a temperature of 90 and fully saturated, each time the air in the hive is changed 1-600 of a pound of water would be removed. A pound of water could be expelled from the hive in 10 hours if the bees could change all the air of the hive once a minute.

When an analysis of the evaporation inside the hive is undertaken we are confronted with many conditions that increase or retard it. Water vapor and warm air being lighter than air at the entrance, pass naturally to the top of the hive, and remain there unless some mechanical force compels them to move toward the entrance. This force is noticable when the hive contains smoke. Under normal conditions we do not find this fanning going on all through the hive.

So far only the evaporation which takes place when the air and water are at the same temperature has been considered. Water can readily be turned into steam by boiling. This steam begins to rise from the surface of the water long before the water comes to the boiling point. "The surface of any watery liquid, whose temperature is 20 degrees warmer than any superincumbent air, rapidly gives off true steam,"* applies to the evaporation in the hive. We know the temperature of the bee's body,

———

*Wells' Natural Philosophy.

and it naturally follows that the temperature of the nectar in the bee's stomach is the same. Observation tells us that the bees ripen the honey drop by drop, bringing it from the stomach and pulsating each drop in the mouth for about ten minutes.‡ The drop during this process is exposed to the air at the particular temperature of that part of the hive where the bee is doing this work. When this temperature is 20 degrees colder than the drop of nectar true steam is given off. This evaporation takes place, no matter how much water vapor is in the air, and varies with the difference of the temperature of the air surrounding the drop of nectar. This is forced evaporation. We find conditions favorable to forced evaporation in cool night temperatures and a hive with a generous around of room.

To clearly comprehend the great amount of water that must be evaporated to get a pound of ripened honey, let us assume that 12 pounds is the weight of one gallon of honey, and 48 pounds the weight of 6 gallons of water. 6 gallons of water mixed with 1 gallon of honey gives 60 pounds of water and honey. 20% of ripened honey is water. 20% of 12 pounds or 1 gallon of honey is 2.4 pounds. 48 pounds plus 2.4 pounds is 50.4 pounds, which is the weight of water in this mixture. 50.4 pounds is 84 % of 60 pounds. 84% is a conservative estimate of the amount of water in nectar. To get 1 gallon of honey 48 pounds of water must be separated from the nectar. To get 1 pound 1-12 of 48 pounds of water must be separated, or 4 pounds. Let us grant that 2 pounds of this water is disposed of before the bees enter the hive. They will still have 2 pounds of water to evaporate before they produce 1 pound of ripened honey. When the honey increase of a hive is 4 pounds a day, 8 pounds of water must be evaporated. The nectar carried into the hive that day would require about three frames for temporary storage. When conditions are such as to hamper evaporation for two or three days of successful honey gathering, it can readily be seen that a crowed condition could exist.

‡A. C. Miller.

NECTAR IN THE FLOWERS

Another way that evaporation influences the work of the bees is that which takes place from the surface of the nectar while it is yet in the flower. Very little is known of the exact percentage of water in nectar. It has been weighed and found range from 60% to 93%. Scientists can never give us much information on this, unless they take into account the evaporation which takes place from the surface of the nectar in the open flowers. Nectar in the same blossom may contain 90% of water at 9 A. M. and 60% at 2 P. M., because of the movements of the air, sunshine and low relative humidity. Because of this variation of the volume of water in nectar, we can understand what a differance there would be in hive conditions, as the nectar is gathered at a time when there is little or much evaporation taking place in the field.

Consider the gathering of nectar from buckwheat. There are districts where the bees work wholly on buckwheat during its blooming season, and peculiarities of the flow of this nectar have been commented on by many beekeepers. Some have noted that the nectar has all disappeared about 2 P. M. They have not gone so far as to prove this by an examination of the flowers, but have assumed it to be so, because the bees did not work on the flowers after this hour. This reasoning is right as to the absence of nectar after this hour.

Other things commented on, are that the bees gather honey from buckwheat only during sunshine periods, and that they gather more on a still day than a windy one.

Let us now assume that all the blossoms that receive the direct rays of the sun secrete nectar suitable for the bees. We have no reason to doubt that flowers on the same plant, surrounded by like conditions viz; sunshine and air, secrete nectar that the bees can gather. The bees can not gather all the nectar from every flower which contains it, before 2 P. M. It is not reasonable to beleive that the flower absorbs what the bees don't gather by this time, and yet we know that it has disappeared. The only way to explain this absence of nectar after this hour is that it has evaporated. Note the statement that the wind reduces the amount of honey gathered. Wind increases evaporation.

Unless some close observer along this line can prove otherwise, we must reason that the flower containing nectar in the morning, and if this nectar has not been removed by any insect, would still contain nectar in the afternoon, so that the bees could gather it, if evaporation had not reduced it to such consistancy as to be useless to the bees.

ABSORBTION OF WATER BY HONEY

It would be impossible for the bees, except under the most favorable conditions, as regards humidity of the air, to produce well ripened honey wholly by evaporation and depending on v ntilation alone to carry off the water vapor. Honey put up in barrels that are not dry, absorb water from the wood, shrinking it enough to cause leakage. One of the causes of watery cappings is that the honey has absorbed water from the moisture-laden air.

Even if the bees ripened all the honey by some process within themselves, as some have suggested, (this would have to allow for only worker-bees doing the ripening and throwing off the water in the flight from the hive), the moisture-laden air, would soon undo the work of the bees. Hence to produce well ripened honey we have to let the bees condense the moisture of the hive on the inside walls and ceilings. An exception of this would be in localities having a very dry atmosphere, but these are very scarce during the urge period.

CONDENSATION

A good understanding of the condensing properties of a hive will make clear a great deal that puzzels us about bees, and such knowledge constructively applied to honey production, cannot fail to give results, the magnitude of which, the beekeeper with his present knowledge, would hardly dare to hope for.

We have seen that ventilation, by moving the warm air out of the hive reduces the heat of the hive. When better results can be obtained in a way that conserves the heat of a hive, a step forward has been taken. This will result if we let the bees, as they are striving to do, use their hive as a condenser.

To understand how water vapor is condensed in a hive, the dewpoint should be understood. Saturated air is at the dewpoint. The amount of moisture that the air will hold varies as the temperature changes. The amount of moisture in the air is spoken of as absolute humidity. Air at O degrees F. is saturated with 1-2 grain, at 60 degrees F. 5 grains, and at 80 degrees F. with 11 grains of water vapor. Air at 60 degrees which contains but 2 1-2 grains of gater vapor would be at 1-2 or 50 per cent of saturation. Air at 78 degrees which contains the same amount of water vapor (2 1-2 grains) would be but 1-4 or 25 per cent saturated. This per cent of saturation is called relative humidity. If the amount of water vapor in the air remains constant as the temperature changes, the relative humidtiy changes with the temperature. Air that is saturated at 78 degrees would have to give up 50 per cent of its water vapor when its temperature is reduced to 60 degrees, leaving this 60 degree air also saturated.

The dewpoint is the temperature of the air, at which it would be at saturation, if the amount of moisture is neither increased nor deminished.

To illustrathe the principle of the air giving up part of its water vapor when its temperature is reduced to the dewpoint, saturate a sponge with water, then press slightly on it and it looses part of its water. The pressure of the hands acts on the sponge the same as reduced temperature on saturated air.

The most obvious illustration of this formation of dew, or condensation of moisture on a cool surface, is when a pitcher of ice water is placed in a warm room. This everybody is familiar with. Condensed vapor on the windows, when the outside temperature is cool, is another. Place a pan, bottom up on the ground and dew will be found on the inside surface of the pan in the morning.

The beehive, when we allow the bees to arrange their own ventilation, condenses vapor this way, when there is enough difference between outside and inside temperature to give a dew-point on the inside surface of the hive. We do not observe any great amount of moisture on the inside walls of the hive, during the urge period, because the bees, needing a good supply of water to mix with the food they prepare, gather it as soon as deposited, leaving the surface in better condition for more condensation. This gives the bees nice distilled water for their use.

Records of weights taken morning and evening, during the honey flow, show but a slight loss in weight of the hive each morning. This has puzzled many beekeepers, and some have suggested that the bees got rid of the water of the nectar before entering the hive.

Let us look at this and apply the condensation theory. The bees evaporate the water from the nectar, condense it on the inside walls of the hive, then gather it as distilled water and mix it with food for the brood. The water has been evaporated, but the hive still retains it, and the hive weights have decreased but little during the night.

We have all had some experience with dew and have observed that all objects will not have dew on their surfaces in the morning, after a night favorable to the deposit of dew. The grass will be very wet, but the road is dry. A painted board is wetter than an unpainted one. The grass and the road are in contact with air containing the same amount of water vapor. Any Natural Philosophy will explain this by the laws governing the radiation of heat. The surfaces of objects on which the dew has been deposited radiate more heat into space, and consequently cool more than the ones that radiate less heat and have no dew on them. The air which comes in contact with the colder surface, and whose temperature is reduced to the dewpoint, is compelled to give up some of its moisture.

The law of the dewpoint operates in the hive the same as outside. Air in the hive at a temperature of 90 degrees which contains 8 grains of water vapor must, when it comes in contact with the cool walls of the hive whose temperature is 60 degrees, give up 3 grains of its water vapor. This condensation will go on as long as temperature of the hive walls, temperature of air in the hive, and humidity of hive air, hold such relations to each other that the dewpoint obtains at the inside surface of the hive.

All objects do not radiate heat equally well. Some part with their heat or become cold faster than others. A smooth stone, exposed the same as grass surrounding it, does not part with its heat as readily as the grass.

PROPOLIS

Blackened tin has a high radiating power and is taken as a standard for expressing the radiating power of other substances. Give the radiating power of blackened tin as 100, rosin is 96 and wood is very low. Propolis, being a rosin its radiating power is almost perfect. But few substances approach this high standard. It is obvious that more condensation takes place on a hive lined with propolis than one where the wood is without a varnish. More condensation makes more evaporation, more exaporation makes more room.

A substance like propolis that is a good radiator and absorber, is a poor transmitter of heat. Propolis used as a varnish keeps the hive heat from warming the wooden walls, and they being cooler at night, give a temperature to the inside surface, that condenses the water vapor fast or slow, as the difference of inside air temperature and wall temperature is great or small.

The bees finding it necessary to separate a great deal of water from the nectar, dispose of it by evaporation, but instinct tells them that the air holds very little on cool nights in May and June; so to remove this moisture from the air, they varnish the inside walls and make a condenser out of the hive. The process is, in principle, the same as that used in distillation.

The makers of refrigerators take advantage of condensation to maintain dry air in the food chamber. A refrigerator car, after years of experimenting has been finished off with several coats of varnish. It has been proven, by tests, that products that are inclined to sweat or absorb moisture, keep drier in a highly varnished car than in one not varnished. The water vapor in the air condenses on the inner surface of the car, making the air drier, consequently preventing the absorbtion of water from the air by the products stored in the car.

SHADE

Shade in the afternoon has been observed to have a tendency to prevent swarming. There can be no condensation if the temperature of the hive walls is nearly equal to or higher than the hive air. This is the condition that exists most of the day and evening when the weather is warm and the sun is shining; providing the hives are not shaded. When a hive is so situated that the shade covers it about noon or soon thereafter, the conditions for condensation are good, cool walls.

Bees situated in dense shade have been observed to swarm more than those having more open surroundings. In dense shade the humidity is high and no movement of air takes place. The hive may be well ventilated as the beekeeper has had best results from well ventilated hives; but the air, being almost saturated when it enters the hive, does not carry off the water vapor from the hive. The good ventilation keeps the inside and outside walls of the hive at nearly the same temperature and if the hives are not well lined with propolis, not much condensation takes place. With nectar not ripening into honey, they soon feel their crowded condition.

CLUSTERING OUT

When the bees are clustering out, it is not the heat of the hive that bothers them, but the high dewpoint in the hive. Man can stand a temperature in dry air much above 100 degrees without much discomforture, but let the air become so saturated with moisture that the dewpoint approaches the temperature of his body, then he feels more or less sufficating effects. We have all noticed this condition of atmosphere on a day that we call close and stuffy. The bees that are clustering out have this close stuffy air in the hive and cluster out for relief.

Cover the outside of the hive with a wet cloth and soon the

bees will be back in the hive about their work. The hive walls are cooled by the evaporation of the water from the cloth, bringing on condensation of hive moisture, consequently lowering the dewpoint.

ABSCONDING SWARMS

Many beekeepers report swarms leaving a short time after being hived. When they have mentioned anything about conditions, a high temperature and a new hive are noted. These make just the right combination to give a high dewpoint in the hive. The parts of a new hive naturally would fit close, and wood being a very poor radiator of heat, no condensation takes place. The swarm objecting to the close stuffy air, and having no home ties, leave for a more suitable place. The remedy for bees that will not stay hived is a few days in the cellar. The cellar temperature is enough colder than the hive a'r to condense the hive moisture on the wood.

WILD BEES

Bees in selecting a home for themselves in a tree, consider two things upon which their future welfare depends, and in doing this we can believe that their instinct led them right. First, a small entrance so that they could easily protect themselves from their enemies. Second, a cavity in the tree of sufficient capacity to carry on the work of the hive, and small enough to keep themselves and their brood from being harmed by outside weather conditions. Never do we find them, in the selection of a home, paying the least attention to ventilation. Natural ventilation is a thing that the bees ignore. Every beekeeper has visible evidence of this truth in his aipary and it cannot be made more clear by further discussion.

LATENT HEAT

Evaporating by the aid of condensation has the advantage that it retains the latent heat in the hive. This might be considered too small a thing to claim our attention. Most things pertaining to a colony of bees are small, but the result of many small things working in a single direction and having the same ultimate goal, is great efficiency. But this is not as small as it appears to be. During the urge period, it conserves great quantities of heat for the colony. The bees' instinct leads them to give consideration to things that make for economy. If one pound of water is evaporated in a hive, the water vapor into which this volume of water has been converted, contains latent heat enough to raise 1,000 pounds (about four barrels) one degree. In changing this back from a gaseous to a liquid state, the same amount of heat is given off in the hive. By ventilation it is lost.

HIVE WALL TEMPERATURE

If a hive of bees be covered with heating-manure to keep it warm in spring the colony will dwindle and become very weak. Hive walls are warmer than hive air and no condensation takes place. For the same reason bees in a greenhouse dwindle. They should be set where the hive walls can cool during the night.

In some loclaities in spring, occurs a succession of days during which the bees do not fly. They have young brood in the hive and when they do not get condensed water from the walls the increase is very slow. A colony of bees, small in proportion to the hive, does not build up as fast as one that is strong for the size of the hive. When bees are not able to heat the hive air above the temperature of the walls, no condensation takes place, and when the bees cannot fly no water is available for domestic purposes. How many beekeepers have stopped to consider where the bees got the water to rear brood before the first flight? Distilation tells the story.

Condensation tells many a story to explain bee behavior. Can you wonder at the awe and amazement with which I approached this phase of bee economy, when brought to a realization of the wonderful instinct that led the bee into the realms of physics, to select propolis, one of the best substances nature has, to bring their work up to a high state of efficiency. This instinct tells them that freer evaporation obtains when the water vapor is being freely condensed on the inner walls of the hive. It tells them that water can be evaporated in any air with a temperature 20 degrees lower than that liquid. It tells them that this water can be condensed readily on the propolis covered walls of the hive. Wonderful is the instinct of the bee!

CONSTRUCTIVE HIVE

The hive that condenses the most water vapor and leaves it so that the bees can dispose of it, is the hive in which we find the contented bees. This water vapor may be allowed to pass from the top of the hive and be condensed the same as in the condensing chamber of a still, but by this process the bees are deprived of the water the same as when the honey is ripened wholly by evaporation. It has been shown in the chapter on evaporation that for every pound of honey ripened in the hive the bees, when they use the condensing system, have for their domestictic use from two to four pounds of water. This amount of water is from 60 to 80 per cent of what nectar the bees brought in that day. Now when this water escapes from the hive and they have to make special trips for water to take the place of this that has escaped, it is obvious that the bees that have to go for water, will reduce the number of nectar gathering bees. This percentage can only be speculated upon, but it runs from 20 to 60 per cent of the field force. When all the field force presist in gathering nectar, as I am almost certain they do at times, the brood is restricted.

What the bees want is a hive that is a good condenser and that retains the condensed water on the inner surface of the hive. The nurse bees can then gather it as they want it. We will have to give the bees credit for being efficient, and it follows that they do not evaporate the water from the nectar that they mix with the food for the brood, but from the surplus. But there occurs a day or a succession of days, during the urge period, when no nectar nor water are brought in, and these are the days when condensed water from a humid atmosphere brings contentment to the colony.

The hive, that from years of experimenting and close observations of weather and hive conditions, gives these results,

I will now try to describe. All my hives are factory made, ten-frame Langstroth dimensions. All flat covers have been set aside and telescope covers substituted. No inner covers, or honey-boards, are used. This cover rests on a rabited cleat that extends all around the hive. This cleat is fastened at a proper distance from the top of the hive-body to allow a beespace between the upper edge of the hive-body and the cover. On the inside top of the cover is fastened a piece of "wall board," 13 3-4 by 17 1-2 inches. This "wall board" is between 1-4 and 3-8 of an inch thick. The cleat on which the cover rests is rabited so as to leave a 3-8 inch space all raound between the hivebody and cover, and over the edges of the hivebody and above the frames. The cover, inside measure, for a hivebody 16 inches wide is 5 1-8 by 16 3-4 by 20 3-4 inches.

All joints and cracks are filled with hot rosin or pitch and the inside of the hivebody and cover is given three coats of varnish.

The cross section of the Constructive Hive given here shows at a glance the consrtuction of the cover and the way it is supported on the hive. Note how similar in principle is this cover to the inverted pan that is set in the garden and which has dew deposited on the inside surface in the morning.

The bees seem to regard this cover as the final word on inside finish, and they are the ones whose approval we must look for. Never have I found one attempt to intrefere on their part. The cover in the fall is just as free from propolis as it was the day it was given to them.

This cover was built as the result of conclusions arrived at from a study of condensation of moisture. The only object when designing it was to make the hive a condenser that would leave the condensed water where the bees could use it, but its good points do not stop at that one thing, as I have observed after its use for some time. These good points are as follows:

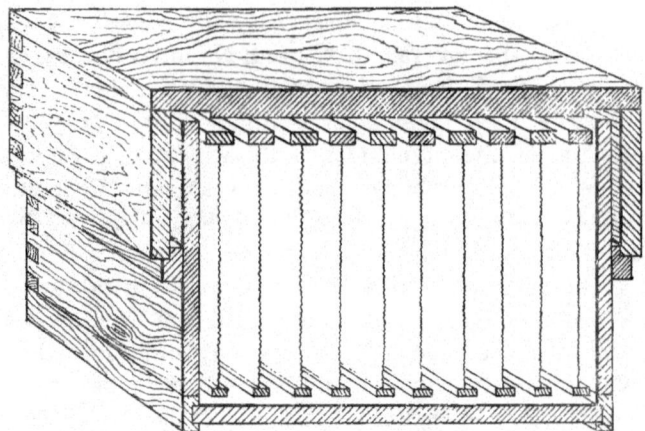

CROSS SECTION OF THE CONSTRUCTIVE HIVE

1. Good condenser.

2. Wind does not blow it off.

3. When once on a bump or jar from person or object does not misplace it.

4. Bees are more gentle to handle.

5. No bees are crushed when putting on the cover.

6. When the cover is removed in cold weather and put on again, the hive is just as comfortable as it was before the cover was taken off.

In regard t othe bees being more gentle to handle, it must be stated that the bees are so gentle as to make this point very noticeable, where Seal-Tight covers are used. This must be accounted for by the reasoning that the bees are contented, and in our not having to destroy their gluing when removing the cover. They are relieved of all work for the prevention of upward ventilation, and the old guers, who are so belligerent are not there when the cover is taken off.

WINTERING

So much has been written on wintering bees that it is not necessary to try the reader's patience by restating what has already been proven of the conditions that give best results.

Just an enumeration:

Outdoor Wintering:

Strong in your bees. Sufficient honey of good quality. No disturbance. Entrance and hive in proportion to the number of bees. Hive. Sufficient insulation. Sealed covers. Absorbent cushions?

Inside Wintering:

Strong in young bees. Sufficient honey of good quality. No disturbance. Large entrance. Ventilation of room good. Temperature of room 57 degrees.

Poor wintering is due to disturbance. Disturbance can be diagramed as follows:

Jaring.			Temperature.	High: Disturbance.
Light.				57 deg. F.:Comfortable.
Noise.				Low: Disturbance.
Poor honey.		Inside the hive		
Temperature of air surrounding the hive.			Humidity.	High: Disturbance.
Humidity of air surrounding the hive.				70 to 80% Comfortable.
				Low: Disturbance.

The bee is so constructed that the temperature of its body can be reduced to 57 degrees.* This reduction of body-temperature the bee does not control, but it is the result of reduced temperature of the air surrounding the bee. The tmperature of the bodyof the bee can be raised in two ways; first, by the temperature of the air surrounding it; second, by a power inherrent in the bee. The bee cannot of his own volition lower his temperature below that of the surrounding air, hence, to attain that

*U. S. Farmers Bulletin, 695.

condition of body that gives winter's quiet and contentment, it depends wholly upon the temperature of the air surrounding it.

This minimum body-temperautre has been found to be 57 degrees. When the temperature of the air surrounding the bee is below this point energy is used by the bee to keep its temperature up to this safe minimum .

The instinct of the bee leads it to do things, not wholly for his own individual comfort, but for the good of the colony. For this reason we find the bees clustered when the temperature of the hive air goes below 57 degrees. The temperature of the cluster may be raised at the center to 90 degrees, the ones at the outside of the cluster are giving their bodies for isolation of the cluster. It is not known if a temperature down to freezing injures these exposed bees. Bees have been revived after exposure, for some time to a very low temperature.

If the reader will keep in mind this relation of temperature to the bees, the effect of humidity of the air on a colony of bees will be better understood. It is evident to every person wintering bees that humidity plays an important part in success or failure. Temperature and humidity are so related to each other that temperature controls the absolute humidity of the air at saturation, as the following tabulation shows:

Temperature	Vapor wt. per Cu Ft. at saturation. Grains.
0	0.54
10	0.84
20	1.30
30	1.97
40	2.86
50	4.09
60	5.75
70	7.99
80	10.95
90	14.81
100	19.79

Where we find a colony with diarrhea we find even the frames and combs damp. This is the condition of most colonies that are dead or very weak in spring. Honey has an absorbing power for water and in a moisture-laden air will absorb great quantities of it. When honey gets thin and watery it is not good feed for bees, and with such a diet the system soon clogs, and we are all familiar with the results.

We naturally ask the cause of this moisture-laden air. Most beekeepers give respiration of the bees as the cause. One day in the fall I took a Seal-Tight cover from the hive and held it so that the drops of water that were condensed on its inside surface could run down to a corner of the cover. Then I poured out a little more than two ounces of water. What yet remained on the cover I estimated to be one ounce. Three ounces of water condensed on the cover. No estimate was made of the amount on the hive-walls. Did the bees breathe out that much water the previous night?

One of the axioms of physics is that something cannot be produced from nothing. If bees are confined in winter quarters 120 days and they consume 30 pounds of honey, of which 1-5 is water, making the consumption of 6 pounds of water during the winter, or 1 1-4 of an ounce for one day. Could they breathe out all the water that they eat, this amount falls away short of the amount that condenses on a Seal-Tight cover in a single-wall hive on a cool night.

OUTDOOR WINTERING

The ordinary beekeeper cannot keep his bees at a uniform temperature; neither can he control the humidity of the air. So his bees have a fluctuating temperature to overcome and consequently a changing humidity. This changing humidity is the one great unseen cause of disturbance in bees.

A clear understanding of the laws of meteorology that relate to moisture in the air, will be necessary to comprehend where this water vapor comes from. The amount of water vapor in the atmosphere is ascertained by using two tested thermometers. One, called the dry bulb, is exposed and the temperature noted. The other one has a cloth covering the bulb. This cloth is moistened and the thermometer whirled in the air for a short time, and the temperature noted. Then by subtracting the temperature of the wet bulb from that of the dry bulb, we have a basis for computing the dewpoint, and from this the relative humidity. Knowing the tmeperature of the dewpoint and the amount of water vapor that saturates the air at that temperature, we have the absolute humidity. Divide the absolute humidity of the dewpoint by the absolute humidity of the dry bulb temperature and you have the relative humidity.

Air can be supersaturated; that is it can contain water in excess of that which saturates it at its temperature, but this excess water is held in the air in globules of water and is known as fog or clouds. The excess water has given up all its gaseous properties and is not subject to the same laws as the water vapor cf the air.

The gaseous vapor of the air is controlled by the same laws, in regard to its expansion and contraction as air. It is well known that if a chamber, in which the air is rairified, is opened, the air rushes in from the outside until the outside and inside air have the same density.

Water vapor does not rush into a hive, where the hive air is relatively dry and a small opening is maintained, but is slowly forced in until its tention at hive temperature equals the tention of the outside vapor at outside temperature. Because of this slow movement of water vapor, when its tention inside the hive is greater than on the outside it condenses on the inner walls, or is transformed into fog, which then looses the expansive power of a gas and is retained in the hive.

Disturbance inside the hive which causes the bees to raise the temperature of the hive air, with its then greater capacity for moisture, or a fall of temperature or raise of relative humidity outside, give an unequal vapor tention and cause vapor to pass from the outside into the hive.

When the hive air is cooling it loses some of its moisture which is condensed in the hive. To illustrate; assume a hive temperature of 57 degrees and an absolute humidity of four grains, which temperature is raised to 70 degrees, and still 4 grains of moisture, to equalize the tention of the vapor, inside and outside, some moisture is slowly forced through the opening; let us say until this 70 degree air holds 6 grains. When this air is reduced to 57 degrees and an absolute humidity of 4 grains, two grains of water is left in the hive after this disturbance.

Suppose the hive is not insulated (no thickness of insolation is too much in any locality whose temperature falls below 50 degrees) and the outside temperature drops to 32 degrees, the hive temperature drops from 57 degrees, humidity 4 grains, to 40 dgrees; at 40 degrees air is saturated with 2.86 grains. In this case about 1 1-2 grains of water is left in the hive.

Let us pass to the next cause, the change of the humidity of the outside air. Assume a maximum temperature of 40 degrees, a relative humidity of 84 degrees and an absolute humidity of about 2.30 grains: the hive air at 50 degrees, a relative humidity of 74 per cent and an absolute humidity of 3 grains. To equalize the tension of the vapor inside and outside the hive a constant movement of vapor into the hive is taking place, which when it is equal to the humidity that the condensing surface of the hive can maintain, is being constantly condensed on the inner surface of the hive. Consider the amount of vapor that is constantly passing into the hive on days when the relative humidity is 96 per cent. There are many days in the more southerly states when a 100 per cent relative humidity obtains.

Because of the small amount of moisture with which a zero atmosphere is saturated (0.54 grains) bees in the north have less moisture to contend with than in the south, where the temperature ranges from 32 degrees up to 60 and 70 degrees. In the latter locality absolute humidity of 32 degree air at saturation is 2 grains; at 60 degrees it is 5.76 grains. In the colder localities, with a temperature of zero or lower for weeks at a time, insulation is necessary to protect the bees from the cold, and in spring and fall to protect them from excess of moisture. In localities where the cold is not so severe but that the bees can beat back what little cold comes in at the entrance, insulation is necessary to keep the hive dry. In the north we insulate to keep the hive warm and dry. In the south you should insulate to keep the hive dry.

The question naturally arises as to how insulation keeps the hive dry. How it keeps the hive warm is obvious. Because atmospheric vapor and temperature are so correlated they must be considered together. Heat is communicated in three ways: by conduction, when it travels from partical to partical in the substance heated; by convection, when the particals of the substance heated move away from the sourre of heat; by radiation, when heat travels through space in all directions from the heated substance.

Insulation has to do with heat communicated by conduction. A substance that is a poor conductor is a good insulator. The escape of heat from a substance or space depends on the amount and quality of insulation surrounding the substance or space. The fireless cooker is a good example of what insulation does.

The wood of which a hive is made is the insulation that surrounds the hive air. Add to this insulation a uniform thickness of leaves, sawdust, chaff or any other insulator and you make it harder for the heat to pass out of the hive by conduction. Hence a well insulated hive has a temperature more nearly uniform than a single wall one.

The temperature of a hive could easily be kept uniform if the bees did not have to have air. An opening must be left for ventilation, and because of this heat is passing out of the hive by convection. Not much heat passes through this opening by radiation or conduction, as air absorbs heat slowly and does not readily part with it. The escape of heat where the opening is not too large, is not great and the radiation and convection of heat from the bees will balance that, where outside temperature is not very low. Where they have both opening and single walled hives (poor insulation) to contend with, they exert energy, and the consequent consumption of stores, to supply the loss.

The more insulation and the greater the number of bees in the hive, the more the temperature of the hive air lags behind as the outside temperature changes. The hive vapor under these conditions moves out slowly and does not saturate the hive air.

When a colony of bees cluster they create for themselves a hive within a hive. The insulation of the cluster hive is the bodies of the bees on the outside of the cluster, and the colder the hive air the better they make this insulation.

Let us now indulge in a little speculation. It is an established fact that the heat of the inside of the cluster increases as the hive temperature falls and vice versa. The honey stores are mostly outside of the cluster and exposed to the vapor of the hive. We know that the bees have ways of controling the vapor of the hive in summer. May it not be possible that the rise in temperature of the cluster is for the purpose of increasing the lag in temperature? This may mean a great deal to them by protecting their honey from moisture.

ABSORBENT CUSHIONS

As understood by beekeepers any porous material placed between the cover and frames for insulation is an absorbent cushion. Claim is made that moisture escapes through this cushion. Sometimes the moisture has tried to escape this way and in the spring the cushion is wet. Sometimes the cushion does not fit snug against the super and an opening is left which gives upward ventilation. Moisture here travels in the path of least resistance and posses off through the largest opening it can find. If a close examination is made in the winter frost will be found where the moisture is passing out and the cushion found dry. Sometimes the cushions are found dry where the burlap is laid across the top of the hive and the super set on this and filled with padking. If this burlap is examined a great deal of propolis is found on it, put there by the bees to prevent upward ventilation. Sometimes this cushion is found dry because the inside of the hive is well varnished with propolis, the insulation is good, which conditions keep the hive air at a proper humdty.

The temperature of an nsulated wall, which has an unequal temperature on its opposite sides, will at the points within this wall be influenced by the distance that these points are from the surface. If the outside temperature is 10 below zero the first quarter of an inch on the outside part of this insulation will be about 10 below zero. The inside surface would be about hive temperature. Unless the laws of nature are bent or broken, water vapor would not travel far in this material, under these conditions, without being condensed.

In winter as well as summer, the big part that a well varnished hive and the consequent condensation play can be seen. Most beekeepers winter their bees in hives used a year or more. These hives give results in proportion to the way they are varnished with propolis and sealed at the top. When all hives are well varnished inside, a seal-tight cover provided, and other necessary things pertaining to good wintering, such as bees, food, sheltered location, insulation and a clear entrance, bee conservation will have advanced.

INDOOR WINTERING

Indoor wintering differs from outdoor wintering in that the repository where the hives are stored is the regulator of the temperature and moisture, instead of each individual hive. A good cellar or other repository where a nearly uniform temperature of 50 to 57 degrees is maintained, and well ventilated, is an ideal place to winter bees.

When setting the hives in the cellar remove the bottom board from each hive; then forget that the bees are in individual hives, and think of your cellar as one big complex hive. About the only use there is for a hive in the cellar is that it is a convenient place to hang the frames. If the temperature and ventilation are good, and the cellar dark, the bees will wniter as well if the frames are taken out of the hive and hung on a rack, and properly spaced.

Small entrances are the primary cause of more dead bees than anything else. Condensation takes place in a hive with a small entrance for the same reason as given in outdoor wintering; a change of hive temperature caused by some disturbance. The botton board must be left off to eliminate this moisture trouble. This gives an equal temperature to hive walls and hive air, and makes condensation on the hive walls impossible. If a colony of bees give off much moisture in respiration, and there is no doubt that they give off some, it is easily diffused through the air by means of the large opening at the bottom.

Let us remember that what we are trying to prevent is excess of moisture in the hive air, and that we control it by our control of the temperature of the repository. Temperature and its relation to moisture is the keystone to successful wintering.

CONCLUSION

When we compare evaporation by the aid of ventilation with that which takes place aided by condensation, and give this an application of the laws of heat, with its three ways of communication; conduction, convection and radiation, the tension of vapors; and the stillness, dryness and density of the atmosphere, our conclusion must be that condensation is so uniform in its results, that it elimniates everything ascribed to locality, but the number of flowers and the weather conditions that affect the flight of bees and the flow of nectar. All other conditions, by the aid of condensation, can be controled by the beekeeper.

Ventilation and shade each make more room in the hive, but not with uniformity under all conditions. So we must add to the treatments we give the bees, a well varnished inner surface to the hive, and a cover that, at no time, permits of upward ventilation. Then the bees will be able to keep the nectar out of the way of a queen, whose egg-laying capacity is increasing daily.

Room, and the procedure whereby the bees automatically make more room as they need it, is the single thing that we have to consider in urge prevention.

www.ingramcontent.com/pod-product-compliance
Lightning Source LLC
Chambersburg PA
CBHW071004180526
45168CB00003B/1283